Published in 2020 by
Rice Farming TV
POB 116
Gridley, California 95948
United States of America
matthew@ricefarmingtv.com
www.ricefarmingtv.com

This book was produced by Rice Farming TV.

First Edition Hardback
ISBN 978-0-578-61341-3

Printed in China.

Visit www.ricefarmingtv.com and watch how rice is grown.

Daddy's Got Dirt

A California Rice Story

Matthew Sligar

For Elena and Stella.
Keep reading, learning and creating.

Illustrated by
Daniel Vincent

Daddy's got dirt!
He calls it ground.
It's out in the country
not in town.

"The ground is good,"
he says it twice.
Daddy's a farmer.
He grows rice.

He works a ton
and I help too.
Daddy taught me
and I'll teach you.

Tweet! Tweet! Tweet!
all the birds sing.
Days become long—
winter turns to spring.

Now we're ready
to break the dirt.
Daddy starts the tractor
and we get to work.

The tractor pulls
a big red disc,
opens the ground
and the dirt is mixed.

We smooth it out
and make it flat.
Pulling a landplane
is how we do that.

Daddy says,
"It will help the yield."
We've turned the ground
into a field.

Daddy lifts a gate
way down below.
From inside the ditch
water starts to flow.

The field is flooded.
One thing we need.
I ask Daddy,
"Hey, where's the seed?"

Ring! Ring! Ring!
Daddy calls the plane.
From up above
seeds fall like rain.

Now the baby rice
needs love and care.
We protect it from weeds
that also grow there.

The rice grows fast
in the summer heat.
The plant is green.
Isn't that neat?

And the rice has friends
like ducks and geese.
Even bunnies and bees
live together in peace.

The plants grow old
and what a treat—
kernels start to form.
That's what we'll eat.

Soon we'll harvest
every bit of grain.
First though, this water
needs to be drained.

In the fall plants turn gold
and fields go dry.
We're closer to harvest
as the days go by.

Vroom! Vroom! Vroom!
the harvester roars.
"Climb on in!"
Daddy yells from the door.

I like to sit with Daddy
as he cuts the rice.
He likes it too 'cause
I give him advice.

The harvester moves slowly.
We never stop—
separating grain and straw,
collecting our crop.

The grain cart pulls up
to our side.
We unload rice
as we drive.

To the truck
the grain cart goes,
and all that rice
the driver unloads.

It's off to the dryer
when the trucks are full—
then to the mill
to remove the hull.

California
Rice

The rice is packaged
with a pretty label.
It's grown on our farm
and ready for your table.

Chirp! Chirp! Chirp!
is the crickets' tune.
The sun finally sets
and up comes the moon.

So how is rice grown?
Well, now you know.
Daddy parks the harvester
and home we go.

Mommy's at the door
with a hug and a kiss.
"Wash your hands," she says,
"I made your favorite dish."

Daddy and I are hungry
and the kitchen smells so nice.
I love it when Mommy cooks,
especially when there's rice.

Free Downloadable Coloring Book

www.ricefarmingtv.com/californiarice

color and share on @ricefarmingtv

Thank you for your support!

GOLD SPONSORS

SILVER SPONSORS

BRONZE SPONSORS